Measuring Max

by Quinn Douglas
illustrated by Christina Miesen

SCHOOL PUBLISHERS

Printed in China

ISBN 10: 0-15-350644-X
ISBN 13: 978-0-15-350644-4

Ordering Options
ISBN 10: 0-15-350599-0 (Grade 2 On-Level Collection)
ISBN 13: 978-0-15-350599-7 (Grade 2 On-Level Collection)
ISBN 10: 0-15-357825-4 (package of 5)
ISBN 13: 978-0-15-357825-0 (package of 5)

5 6 7 8 9 10   985   15 14 13 12 11 10 09

Max lives next door. He's only
little, and sometimes I help his mom
look after him.

3

Last year, when it was Max's third birthday, I made him a cool present. I got the idea for his present from a store.

I saw a chart there. "I have a question, Mom," I said. "What is that chart for?"

"It's a chart to show how tall
children are," answered Mom. "Stand
against it, Carla, and I'll tell you how
tall *you* are."

5

I decided to make Max a chart for his birthday. That night, I painted zoo animals on the chart. I had to draw them first, and then I painted the colors.

I remember Mom saying, "I sure hope Max won't worry about the lion. It looks so fierce!"

Of course, Max had liked the lion picture best of all!

I glued a measuring tape on the
chart. At his party, Max stood against
the tape. We could see how tall he was.

8

Now, it's a year later, and it is Max's birthday again. Mom and I have been invited to his house for dinner.

I bought Max some paints today,
and I made him a special birthday card.
Mom made a lion birthday cake
especially for his party. She made sure
it did not look too fierce!

Now that we have had our dinner, Max's grandmother brings the cake to the table. We watch as Max blows out the four candles, and I think how tall he looks.

Suddenly I remember Max's chart.
"Wait a minute," I say. "Before we
eat the cake, let's see how tall Max is
now that he's four."

We all go into Max's room, and he stands in front of the chart.

"You are much taller than you were last year!" I say with delight. "You must be eating all your vegetables!"

"We can have vegetables later,"
laughed Max's dad. "First, let's eat
birthday cake!"

14

Think Critically

1. Where did Carla get the idea for Max's third birthday present?

2. How did Carla make Max's chart?

3. If you were Max, how would you have felt when you saw your lion cake?

4. What do you think the author wanted you to learn from Carla?

5. What part of the story did you like the most? Why?

 Visual Arts

Make a Birthday Card Fold a piece of card in half. Decorate the front with a birthday picture. Write a special birthday message inside the card. Give your card to the next person you know who has a birthday.

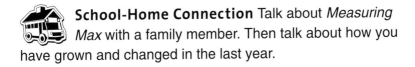 **School-Home Connection** Talk about *Measuring Max* with a family member. Then talk about how you have grown and changed in the last year.

Word Count: 322